Passive Solar Heatin
Buildings

Harlan H. Bengtson, PE, PhD

Emeritus Professor of Civil Engineering
Southern Illinois University Edwardsville

Founder, www.EngineeringExcelTemplates.com

i

Table of Contents

Page

Introduction 1

Chapter 1 – Passive Solar Heating Definition 4

Chapter 2 – Components of a Passive Solar Heating System ... 5

Chapter 3 – Basic Passive Solar Heating System Types 7

Chapter 4 – Inputs Needed to Estimate Size/Performance
                         of a Passive Solar Heating System ... 14

Chapter 5 – Size and Performance Calculations
                         for Passive Solar Systems 25

Chapter 6 – Choice of Type(s) of Passive Solar
                         System to Use 33

Chapter 7 – Sizing Solar Storage 36

Chapter 8 – Summer Shading of Passive Solar Glazing 42

Chapter 9 – NASA/Langley Surface Meteorology
                         and Solar Energy Website 39

References 44

Introduction

The principles of passive solar heating, such as basic types of systems, their description, and the components making up any passive system are presented here. Sources of data for heating requirements and available solar radiation throughout the U.S are identified and discussed along with a method for estimating the rate of heat loss from a home. The use of these three inputs in a method for estimating performance of a passive heating system of specified size at a specified location is presented. The data retrieval and calculations are illustrated with numerous examples.

Passive Solar Exterior

Image Source: https://greenpassivesolar.com/

Passive Solar Interior

Image Source: https://greenpassivesolar.com/

1. Passive Solar Heating Definition

Passive solar heating of a building means using the sun's radiant energy to provide heat by converting the radiant energy to thermal energy (heat) when the building absorbs the radiant energy. Some of the incoming thermal energy may be used to directly heat the building and some may be stored in components of the building. In a completely passive system, energy flow within the building is by natural means (conduction, natural convection and radiation) only. An active solar heating system, on the other hand, uses devices like blowers, pumps and/or fans to move heated fluid, from the collectors to the heated space, from thermal storage to the heated space, and from the collectors to thermal storage.

2. Components of a Passive Solar Heating System

A passive solar system is made up of components that have functions very similar to the components of an active solar heating system, but those components look much different and are arranged much differently. The typical components of either a passive or active solar heating system are: **aperture** (opening for solar radiation to go through, **absorber** (to absorb the radiant energy and convert it to thermal energy, **thermal mass** (for storage of excess thermal energy for later use), **distribution system**, **controls**, and a **backup heating system**

In a passive solar heating system, the aperture and absorber are separated physically, while in an active system they are typically both part of the collectors. The **aperture**(s) in a passive solar heating system will be south-facing window(s). It is important that these windows not be shaded by other buildings or trees from 9:00 a.m. to 3:00 p.m. during the heating season.

In a passive system, the **absorber** and the **thermal mass** for storage are both part of the same unit(s), which are components of the building such as floors and walls in the direct path of sunlight. The floor and/or wall surfaces are typically dark colored so that they will absorbs solar radiation well and convert it to thermal energy, which is stored in the mass of the floors and/or walls. In an active system, the absorber is typically part of the collector and the heat storage system is a separate unit.

A **distribution system** is used to circulate heat from the collection and storage points to different areas of the house. In a strictly passive system, heat will be circulated solely by one or more of the three natural heat flow methods, conduction, natural convection, and radiation. Fans, and/or blowers are sometimes used to help with the distribution of heat throughout the house in an otherwise passive system.

In a passive heating system, **Controls** include items such as operable vents or dampers, moveable window insulation, and/ or roof overhangs or awnings that shade the aperture during summer months. For active systems and nearly passive systems that use fans and/or blowers, controls typically include electronic sensing devices, such as a differential thermostat that signals a fan to turn on or off or a damper to open or close.

The **backup heating system** for either a passive or active system may be any type of non-solar heating system.

3. Basic Passive Solar Heating System Types

The five basic types of passive solar heating systems to be discussed in this book are: **direct gain, thermal storage wall, attached sunspace, thermal storage roof,** and **convective loop.** Each of the types contains the components described above. Three of these types, thermal storage wall, attached sunspace and thermal storage roof, are sometimes called indirect gain systems. The convective loop is sometimes called an isolated gain system. A brief description of each of the five types follows.

Direct Gain is a simple, straightforward approach to passive solar heating. In this type of system, sunlight enters the living space through south facing windows during the daytime. The incoming solar radiation strikes the walls, floor, and/or ceiling (thermal storage mass) of the living space directly or is reflected to them and is absorbed and converted to thermal energy in that thermal storage mass. The heated wall, floor, and/or ceiling will keep the living space warm. Also, during the daytime, when the storage surface temperature is high enough, heat will be conducted from the hot surface of the walls, floor, and/or ceiling to their interior, where it is stored. When the surface of the thermal storage mass are no longer being heated by the sun's rays, the room and the wall, floor and/or surface temperatures will decrease and heat will be conducted from the heated interior of the thermal storage mass to the cooler surface and will heat the living space by convection and radiation. This is illustrated in **Figure 1**, below.

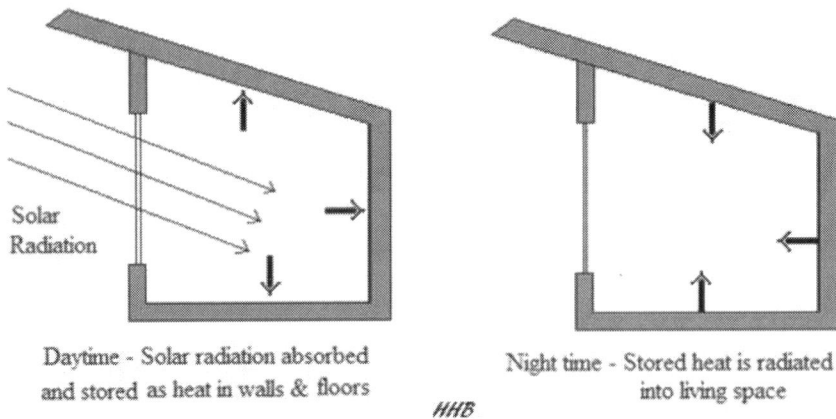

Daytime - Solar radiation absorbed
and stored as heat in walls & floors

HHB

Night time - Stored heat is radiated
into living space

Figure 1. Direct Gain Passive Solar Heating

A direct gain passive solar heating system needs to have
plenty of south facing glass and adequate thermal storage capacity in
the living space. A guideline sometimes used for thermal storage
capacity is to construct one-half to two-thirds of the total interior
surface area of thermal storage materials. Materials typically used
for thermal storage are masonry, such as concrete, adobe, brick, etc.
Also water walls can be used as well. A water wall consists of water
in plastic or metal containers placed in the direct path of the
sunlight. Water walls heat more quickly and more evenly than
masonry, but they may not be as pleasing aesthetically. The surface
temperature of dark colored masonry surfaces may become quite
high if they receive direct sunlight. One way to avoid this, is to use
a diffusing glazing material which scatters sunlight, thus distributing
it more evenly over walls, ceiling, and floor. This approach
distributes the incoming solar radiation more evenly but doesn't
reduce the total amount of solar energy entering the space. The
effect of using a diffusing glazing material is illustrated in **Figure 2**,
below. Nighttime heat loss can be reduced by the use of moveable
insulation, such as insulated drapes, to cover the south facing
windows at night.

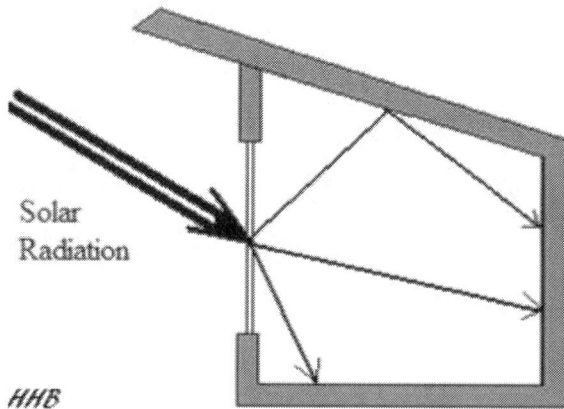

Figure 2. Direct Gain with Diffusing Glazing Material

A **Thermal Storage Wall** is shown in **Figure 3** below. These walls are typically made of masonry or of water filled containers. This type of wall is sometimes called a Trombe Wall. It is named after the engineer, Felix Trombe, who popularized the design together with architect, Jacques Michel, in the 1960's. In 1881, this design was patented by Edward Morse. The key feature of the thermal storage (Trombe) wall system is the presence of thermal storage material between the interior living space and the sun. As a result of this wall placement, the dark colored storage wall will be heated by sunshine during day and the stored heat in the wall will provide heat to the living space at night. Vents are typically placed near the top and the bottom of the wall, as shown in Figure 3. These vents should be opened during the day and closed at night, to provide natural convection* heating of the living space from the heated air between the storage wall and glazing during sunlight hours and to minimize heat loss from the heated space at night. As with the direct gain system, use of movable insulation for the glazing at night will reduce nighttime heat loss.

*NOTE: Natural convection (also called free convection) is the movement of a fluid because of the reduced density of a heated portion of the fluid. That is, as a portion of a fluid is heated, its density decreases and it rises. This will cause movement of other portions of a fluid mass as well. In the thermal storage wall, for example, the heated air between the glazing and the storage wall will rise and enter the living space through the upper vent. This will draw cool air from the bottom of the living area into the space between the glazing and storage wall through the lower vent.

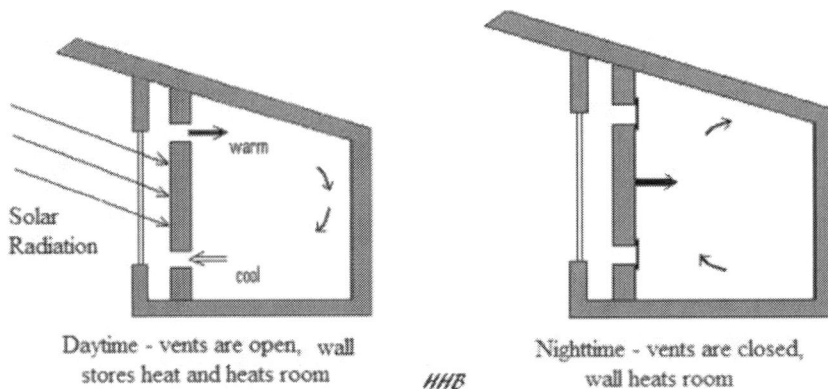

Daytime - vents are open, wall stores heat and heats room

HHB

Nighttime - vents are closed, wall heats room

Figure 3. Thermal Storage (Trombe) Wall, Daytime and Nighttime Configurations

Figure 4 shows an **Attached Sunspace** passive solar heating system. A useful feature of this type of system is that it can be added to an existing building. This makes it more suitable for retrofit than some of the other types of passive heating systems. This type of system uses direct gain in the sunspace, which is sometimes called a solar greenhouse, and may, in fact, be used as a greenhouse. Another important part of the attached sunspace system is a thermal storage wall between the sunspace and the rest of the living space, as shown in Figure 4. Vents are typically included at

10

the top and bottom of the thermal storage wall, as described above for a thermal storage (Trombe) wall.

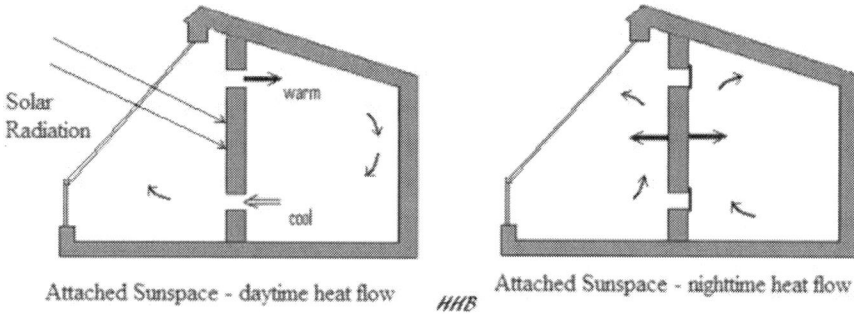

Attached Sunspace - daytime heat flow   *HHB*   Attached Sunspace - nighttime heat flow

Figure 4. Attached Sunspace, Daytime & Nighttime Heat Flows

The **Thermal Storage Roof,** also called a solar roof, solar pond, or roof pond, uses water encased in plastic on the roof. Moveable insulation is used to cover the roof and reduce heat loss during the night, but the roof must be uncovered during the daytime to allow the sunshine to strike the pond. The pond (water encased in plastic) can also be placed in an attic, under glazing in a pitched roof. **Figure 5** shows typical daytime and nighttime heat flows for a thermal storage roof system. On the minus side, however, a thermal storage roof requires a somewhat elaborate drainage system, movable insulation to cover and uncover the water at appropriate times, and a structural system to support up to 65 lbs/sq ft dead load.

11

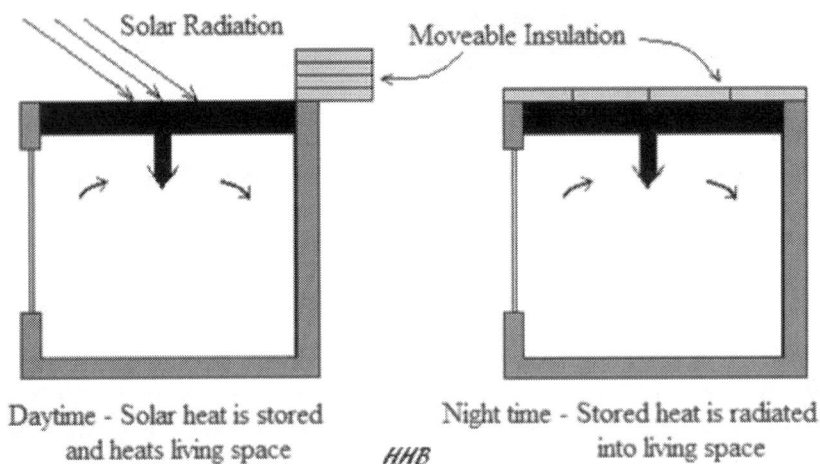

Daytime - Solar heat is stored and heats living space

HHB

Night time - Stored heat is radiated into living space

Figure 5. Solar Roof, Covered with insulation at night, Uncovered during day

A convective loop system uses a flat plate solar collector to heat air or water. The heater air or water then flows by natural convection to directly heat a living space or to a thermal storage area. Two convective loop configurations for space heating are shown in Figure 6. One of the systems has a vertically mounted collector, and the other has the collector mounted at a tilt. Either of these systems can be mounted on an existing wall, so they are quite suitable for retrofit applications. Either of these systems needs two openings into the building, one at the top of the collector for heated air flow into the building, and one at the bottom of the collector for cool air flow from the building into the collector. Some applications use a window on one floor for the upper opening and a window on a lower floor for the bottom opening. The vertically mounted collector, as shown at the right in **Figure 6**, is also referred to as a solar chimney. Other names used for any of the convective loop systems are isolated gain and thermosiphon system. This type of system can also be used for a passive water heating system as shown in **Figure 7**.

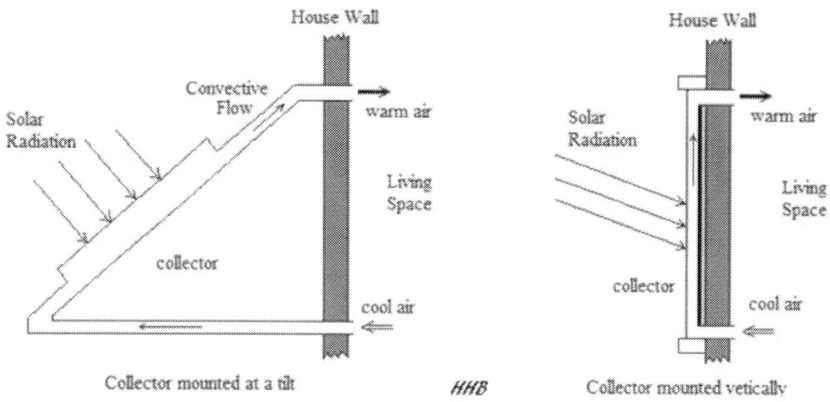

House Wall

Convective Flow

Solar Radiation

warm air

Living Space

collector

cool air

Collector mounted at a tilt

HHB

House Wall

Solar Radiation

warm air

Living Space

collector

cool air

Collector mounted vetically

Figure 6. Convective Loops for Space Heating

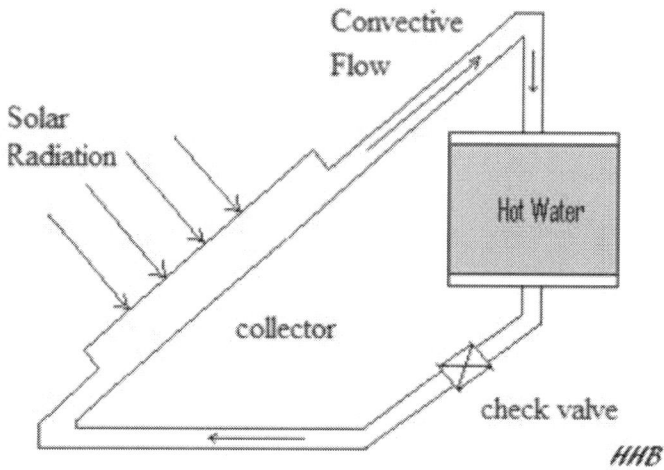

Convective Flow

Solar Radiation

Hot Water

collector

check valve

HHB

Figure 7. Convective Loop for Water Heating

13

4. Inputs Needed to Estimate Size or Performance of a Passive Solar Heating System

In planning and designing a passive solar heating system, it is often helpful to estimate the performance of a particular passive solar heating system, or to estimate the size system needed to provide a specified percentage of the heating requirements for a building. In order to make either of these calculations, information is typically needed on each of the following: i) heating requirements (degree days) during the heating season at the site of interest, ii) information on the rate of heat loss from the house, and iii) available solar radiation at the site of interest. Sources of information for, or means of estimating those three items, will be discussed in this chapter. Then the use of that information for sizing and performance calculations will be covered in the next chapter.

i) Heating requirements (degree days) during the heating season at the site of interest: Data on heating degree days is available from a variety of sources. Two such sources will be discussed here. One very good source of data for passive solar heating applications is *Solar Radiation Data Manual for Buildings,* published by the National Renewable Energy Laboratory (NREL), and available for free download at the website: http://rredc.nrel.gov/solar/pubs/bluebook/. This publication includes data for 239 locations in the United States and its territories, based on data collected from 1961-1990 for those 239 sites. Monthly average and yearly average values for heating degree days are given

14

in the *Solar Radiation Data Manual for Buildings* for each of the 239 sites covered in that publication.

The other source for passive solar heating data that will be discussed in this section, is the second reference in the "Related Links and References" for this course, *Passive Solar Energy: The Homeowners Guide to Natural Heating and Cooling,* by Bruce Anderson and Malcolm Wells. It is available for free download from the website: http://www.builditsolar.com/Projects/SolarHomes/PasSolEnergyBk/ PSEbook.htm.

Appendix 5 of *Passive Solar Energy: The Homeowners Guide to Natural Heating and Cooling*, contains data for 236 U.S. Cities based on ASHRAE, *Handbook of Fundamentals*, 1972. Information is given on monthly average heating degree days for the period from September through May, and the yearly total for each of the 236 U.S. cities.

NOTE: Degree days data provides information on heating or cooling requirements for buildings at a given location. The value of degree days for a given day is defined as the difference between the average temperature for the day (calculated by averaging the maximum and minimum temperature for the day) and a base temperature (typically 65° F or 18.3° C). The difference is called heating degree days if the average temperature is less than the base temperature. It is called cooling degree days if the average is greater than the base temperature.

Example #1: Compare the values given in the above two references discussed above for annual heating degree days and for January heating degree days for Colorado Springs, CO.

Solution: This simply requires looking up values in the tables in those two references.

Table 1 below shows heating degree data and cooling degree data for Colorado Springs CO, from page 47 of the NREL, *Solar Radiation Data Manual for Buildings*.

Colorado Springs, CO	Average Climatic Conditions						
	Jan	Feb	Mar	Apr	May	June	July
Daily Minimum Temp (°F)	16.1	19.3	24.6	33.0	42.1	51.1	57.1
Daily Maximum Temp (°F)	41.1	44.6	50.0	59.8	68.7	79.0	84.4
HDD, Base 65°F	1122	924	859	558	302	87	6
CDD, Base 65°F	0	0	0	0	0	87	186

Colorado Springs, CO	Average Climatic Conditions					
	Aug	Sept	Oct	Nov	Dec	Year
Daily Minimum Temp (°F)	55.2	47.1	36.3	24.9	17.4	35.4
Daily Maximum Temp (°F)	81.3	73.6	63.5	50.7	42.2	61.6
HDD, Base 65°F	18	164	468	816	1091	6415
CDD, Base 65°F	120	26	0	0	0	419

Table 1. HDD and CDD data for Colorado Springs, CO

Source for this data: NREL, *Solar Radiation Data Manual for Buildings*, p 117

From **Table 1** (the NREL manual):

Annual heating degree days = 6415 °F days

January heating degree days = 1122 °F days

16

Table 2 below shows heating degree data for selected Colorado Cities, From Anderson & Wells, *Passive Solar Energy: The Homeowners Guide to Natural Heating and Cooling*, page 155.

Colorado	Heating Degree Days				
	Sept	Oct	Nov	Dec	Jan
Alamosa	279	629	1065	1420.0	1476
Colorado Springs	132	456	825.0	1012	1128
Denver	112	428	819	1035	1132
Grand Junction	10	311	786	1113	1209
Pueblo	54	326	750	986	1085

Colorado	Heating Degree Days				
	Feb	Mar	Apr	May	Year
Alamosa	1162	1029	696	440	8529
Colorado Springs	938.0	891	582	319	6423
Denver	938	887	558	288	6283
Grand Junction	907	729	387	146	5641
Pueblo	871	772	429	174	5462

Table 2. HDD data for Selected Colorado Cities

Source for this data: Anderson & Wells, *Passive Solar Energy: The Homeowners Guide to Natural Heating and Cooling*, page 155.

From **Table 2** (Anderson & Wells book)**:**

Annual heating degree days = 6423 °F days

January heating degree days = 1128 °F days

The agreement between the two sources is quite close, although the data came from databases covering different timeframes.

ii) Information on the rate of heat loss from the house: A useful way of expressing the rate of heat loss is Btu per heating degree-day per square foot of building floor area. For conventional building design, a heating load of 6 to 8 Btu/°F-day/ft^2 is considered an energy conservative design. For an older home that is not well insulated and weather-stripped, the heat loss rate would be greater. For a super-insulated home, it would be less than the above guideline. For an existing home that is not well insulated and weather-stripped, it is typically cost-effective to add insulation and/or weather-stripping before adding passive solar components. Goswami, Krieth, & Krieder (ref #1 at the end of this book) suggest reducing the nonsolar rate of heat loss by 20% in conjunction with solarizing of the south-facing wall of a building with passive solar systems. This gives a range of 4.8 to 6.4 Btu/°F-day/ft^2, as an estimate for the rate of heat loss from a well-insulated and weather-stripped home with passive solar heating system(s) added.

The rate of heat loss per degree-day from a building can be estimated from information about the construction of the building, such as roof area, wall area, insulation thickness, glass area, number of glazing layers, etc. This type of calculation is covered in various HVAC texts and publications and will not be covered here.

Another approach to estimating rate of heat loss from an existing home is through the use of fuel consumption information for that home. Anderson & Wells (ref #2 at the end of this book) discuss this approach on pages 112 & 113 of their book. That discussion is summarized here for several types of heating: fuel oil furnace, natural gas furnace, electric resistance heating and heat pump.

i) Fuel Oil Furnace: A gallon of fuel oil has an energy content of 135,00 to 140,000 Btus. If this is multiplied by the furnace efficiency (typically 40 to 70 %) that gives the heat supplied to the house per gallon of fuel oil used.

ii) **Natural Gas Furnace:** One cubic foot of natural gas contains 1000 Btus of energy. Natural gas consumption is often expressed as hundreds of cubic feet (ccf) or thousands of cubic feet (mcf). As with fuel oil, the amount of energy in the natural gas consumed must be multiplied by the furnace efficiency to get the amount of heat delivered to the house.

iii) **Electric Resistance Heating:** For electric resistance heating, one kwh of electricity is equivalent to 3400 Btu with electric resistance heating having an efficiency of 100%.

iv) **Heat Pump:** A heat pump can supply as much as 6800 Btus per kwh of electricity used to operate it.

This information can be used together with an estimate of the amount of fuel or electricity used to heat the home from power bills, to estimate the amount of heat supplied to the house during any month. By the first law of thermodynamics, that must be the amount of heat lost from the house.

Example #2: Data on monthly natural gas consumption for a 2000 ft^2 home in Albuquerque, NM is given in **Table 3** below. Also given in the table are monthly heating degree-days for Albuquerque, NM from the NREL publication, *Solar Radiation Data Manual for Buildings*. Estimate the rate of heat loss from this home in Btu/°F-day/ft^2 based on this data. Assume 60% efficiency for the furnace.

Month	Jan	Feb	Mar	Apr	May	June	July
Natural Gas Consump., ccf	272	186	95	73	27	15	18
Heating °F-days	955	700	561	301	89	0	0

Month	Aug	Sept	Oct	Nov	Dec	Year
Natural Gas Consump., ccf	17	17	28	91	186	1025
Heating °F-days	0	18	259	621	921	4425

Table 3. Natural Gas Consumption and HDD data for **Example #2**

Solution: Ave. gas consumption for June – August (zero heating degree-day months) is:

$$(15 + 18 + 17)/3 \;=\; 17 \;\underline{\text{ccf/month}}$$

Total gas consumption for heating season (Sept – May):

$$\underline{\mathbf{= 1025 - (15 + 18 + 17) \;=\; 975\ ccf}}$$

Calculate the annual gas consumption for heating by subtracting the baseline   (non-heating) gas consumption of 17 ccf/month:

Gas consumption for heating $= 975 - 9*17 = \underline{822\ ccf}$

Converting to Btu: Gas consumption for heating $= 822*10^5 = 8.22*10^7$ Btu

Heat delivered to house (60% efficiency) $= 0.6*8.22*10^7 = 4.93*10^7$ Btu

Dividing by the average annual heating degree-days for Albuquerque and the floor area of the house:

Heat loss rate in Btu/°F-day/ft^2 $= 4.93*10^7/4425/2000$

$$= \underline{\mathbf{5.6\ Btu/°F\text{-}day/ft^2}}$$

This seems reasonable. It is in the range of 4.8 to 6.4 Btu/°F-day/ft² given above for a house that is receiving some solar input from south-facing windows.

iii) Available solar radiation at the site of interest: Data on incident solar radiation in the United States and around the world are available from various publications and internet websites. The NREL *Solar Radiation Data Manual for Buildings*, that was introduced in the previous section on heating requirement data is especially useful for passive solar heating applications, and will be discussed here.

The *Solar Radiation Data Manual for Buildings* includes data on monthly average and yearly average incident solar radiation on horizontal surfaces and on vertical surfaces facing north, east, south and west. **Table 4** below shows some of the data from page 117 of this reference for Kansas City, MO.

The first part of **Table 4** gives information about incident radiation. The "global" monthly radiation values given in the table are the average total solar radiation striking the surface, including the effect of cloudiness. "Diffuse" monthly radiation is that part of the global radiation that is made up of ground reflected radiation and sky radiation. The remainder of the global monthly radiation is "direct beam" radiation reaching the surface directly from the sun. The "clear day global" radiation rate is for a day with no clouds in the sky at that location.

The second part of **Table 4** gives monthly and yearly average transmitted radiation through double glazing for horizontal and for east facing and south facing vertical surfaces. The "shaded" transmitted radiation is for shading with an awning or overhang designed to shade the window from the sun's rays in the summer months, but not in the winter months. The required dimensions for such shading is discussed in more detail in Chapter 8. The

21

"unshaded" transmitted radiation is for the case with no awning or overhang above the window. The table on page 117 in the NREL manual includes similar data for north facing and west facing windows. The data of primary concern for passive solar heating is that for south facing windows.

Similar data is available for all 239 locations in the United States and its territories that are included in the publication.

Kansas City, MO	Average Incident Solar Radiation (Btu/ft^2/day), Uncertainty ± 9%						
	Jan	Feb	Mar	Apr	May	June	July
Horizontal Global	700	940	1240	1610	1870	2050	2080
Minimum	560	790	980	1330	1560	1750	1800
Maximum	820	1100	1410	1870	2130	2360	2350
Diffuse	300	420	560	680	790	810	760
Clear Day Global	940	1280	1760	2230	2530	2640	2560
East Global	480	610	760	940	1040	1110	1130
Diffuse	240	330	430	520	590	630	620
Clear Day Global	700	900	1140	1350	1450	1470	1440
South Global	1170	1170	1100	990	850	780	840
Diffuse	350	420	490	530	560	560	560
Clear Day Global	1940	1980	1800	1400	1060	910	960
Average Transmitted Solar Radiation (Btu/ft^2/day) for Double Glazing, Uncertainty ± 9%							
	Jan	Feb	Mar	Apr	May	June	July
Horizontal - Unshaded	450	640	880	1160	1350	1490	1520
East Unshaded	339	430	540	620	740	790	810
Shaded	300	380	470	580	630	680	700
South Unshaded	880	850	760	640	520	470	500
Shaded	860	790	610	420	350	350	350

Table 4 – part 1 - Avail. Solar Radiation Data for Kansas City, MO

Source for this data: NREL, *Solar Radiation Data Manual for Buildings*, p 117

Kansas City, MO	Ave Incident Solar Radiation (Btu/ft²/day), Uncertainty ± 9%					
	Aug	Sept	Oct	Nov	Dec	Year
Horizontal Global	1830	1460	1120	740	590	1360
Minimum	1550	1060	930	590	490	1280
Maximum	2040	1760	1390	890	700	1490
Diffuse	680	560	410	320	280	550
Clear Day Global	2290	1870	1390	990	820	1780
East Global	1040	880	730	500	400	800
Diffuse	550	460	350	260	220	430
Clear Day Global	1350	1180	950	740	630	1110
South Global	980	1130	1280	1120	1030	1040
Diffuse	550	500	430	350	330	470
Clear Day Global	1220	1610	1870	1900	1870	1530
Ave Transmitted Solar Radiation (Btu/ft²/day) for Double Glazing, Uncertainty ± 9%						
	Aug	Sept	Oct	Nov	Dec	Year
Horizontal - Unshaded	1330	1050	780	490	380	960
East Unshaded	740	620	510	350	280	570
Shaded	650	550	460	310	250	500
South Unshaded	610	760	920	840	780	590
Shaded	380	550	810	810	760	590

Table 4 – part 2 - Avail. Solar Radiation Data for Kansas City, MO

Source for this data: NREL, *Solar Radiation Data Manual for Buildings*, p 117

Example #3: Find an estimate of the average amount of solar radiation per day (in Btu/day), which will strike a 24" by 36" south facing window in Kansas City, MO, in January.

Solution: From the **Table 4** data extracted from *Solar Radiation Data Manual for Buildings*, page 117, the average daily incident global solar radiation on a south-facing surface in Kansas City, MO in January is 1170 Btu/ft²/day. For the 6 ft² window in the example, the daily incident radiation would thus be: $(1170 \text{ Btu/ft}^2/\text{day})(6 \text{ ft}^2)$ = **7020 Btu/day**

Example #4: From the same source used in **Example #3**, find an estimate of the average amount of solar radiation per day (in Btu/day), which will be transmitted through an unshaded 24" by 36" south facing, double pane, glass window in Kansas City, MO, in January.

Solution: From the Table 3 data extracted from *Solar Radiation Data Manual for Buildings*, page 117, the average daily solar radiation transmitted through an unshaded, south-facing, double pane, glass window, in Kansas City, MO is 880 $\text{Btu/ft}^2/\text{day}$. For the 6 ft^2 window in the example, the daily incident radiation would thus be:

$$(880 \text{ Btu/ft}^2/\text{day})(6 \text{ ft}^2) = \textbf{5280 Btu/day}$$

Note: Based on these figures, the percentage of the incident solar radiation that is transmitted through the double pane window is $(5280/7020)*100\% = 75\%$.

5. Size and Performance Calculations for Passive Solar Systems

The primary parameter to be determined in order to size a passive solar heating system, is the area of south-facing glazing. The other major component, heat storage, then can be sized to match the area of glazing. The fraction of a building's heating load that is handled by solar heating is a parameter often used in some way when designing or sizing a passive solar system. A larger area of glazing, will result in larger solar savings fraction for a given building. Anderson & Wells (ref #2 at the end of this book) provide some rules of thumb and guidelines for preliminary sizing of a passive solar heating system. Calculations to check more closely on the preliminary sizing decision can then be made. That approach will be discussed and illustrated with examples here. Another, more detailed, sophisticated calculation procedure is available in Goswami, Krieth, & Krieder (ref #1 at the end of this book).

Anderson & Wells suggest that the average percentage of daytime hours that the sun is shining during the heating season at the location of interest can be used as an estimate of the optimum percentage of the building heating load to be provided by a passive heating system. The average percentage of daytime hours that the sun is shining during the heating season can be obtained from the maps in Appendix 3 of Anderson & Wells' book. Looking at those maps shows that this percentage is approximately 50% for much of the United States. The fraction is notably higher in the southwest and lower in the northwest.

Anderson & Wells also suggest that on the average each square foot of south-facing glazing will supply about the same amount of heat over a heating season as that delivered by a gallon of fuel oil, about 60,000 Btu. This figure will be lower in very cloudy areas, perhaps as low as half, or 30,000 Btu. In the sunny southwestern U.S., it will be higher, perhaps as high as twice or 120,000 Btu.

Example #5: Using Anderson & Wells' guidelines given above, estimate the approximate glazing area that would be optimum for a passive solar heating system for the Albuquerque house described in **Example #2**.

Solution: From Appendix 3 of Anderson & Wells' book, the average percentage of daytime that the sun is shining in Albuquerque, NM is 60 to 65% for Nov, Dec, Jan, & Feb (most of the heating season). Choose a target of 60% solar reduction in heating fuel consumption. From Anderson & Wells' rules of thumb for Btu/sq. ft. of glazing, try 120,000 Btu/ft^2 as an estimate for the Albuquerque area.

From **Example # 2,** the estimated heat requirement for heating the house in a season is $4.93*10^7$ Btu. The required glazing to deliver 60% of this heat is thus estimated as:

$$(0.6)(\ 4.93*10^7 \ Btu)/120,000 \ Btu/ft^2 \ = \ 246 \ ft^2$$

Example #6: Using data for heating degree days and rate of solar radiation transmitted through south-facing, double layer glazing in Albuquerque, NM, estimate the percentage of the heating load that would be provided by the 246 ft^2 glazing area calculated in **Example #5** for the 2000 ft^2 home in **Example #4**, for each month of the heating season. Use the heat loss rate of 5.6 $Btu/^oF\text{-}day/ft^2$, calculated in **Example #2** for this house.

NOTE: Table 5 below shows the average transmitted solar radiation rate through double glazing by month in Albuquerque, NM and the heating degree days by month in Albuquerque, NM. These figures came from page 148 (the Albuquerque, NM page) in the NREL *Solar Radiation Data Manual for Buildings.*

Albuquerque, New Mexico	Average Transmitted Solar Radiation (Btu/ft²/day) for Double Glazing, Uncertainty ± 9%						
	Jan	Feb	Mar	Apr	May	June	July
South Unshaded	1230	1170	970	720	510	430	460
South Shaded	1220	1110	800	490	360	340	350
HDD, Base 65°F	955	700	561	301	89	0	0

Albuquerque, New Mexico	Average Transmitted Solar Radiation (Btu/ft²/day) for Double Glazing, Uncertainty ± 9%					
	Aug	Sept	Oct	Nov	Dec	Year
South Unshaded	600	860	1140	1230	1210	880
South Shaded	400	650	1040	1200	1200	760
HDD, Base 65°F	0	18	259	621	921	4425

Table 5. Transmitted Solar Radiation Rate and HDD data for Albuquerque, NM

Source for this data: NREL, *Solar Radiation Data Manual for Buildings*, p 148

Solution: The data and calculations are summarized in the spreadsheet table copied below. The second and third columns give monthly values for heating degree-days and solar radiation transmitted through a shaded, double glazed, south-facing window in Albuquerque, NM, from page 148 of the NREL publication, *Solar Radiation Data Manual for Buildings*. (see **Table 4** above). The fourth column was calculated by multiplying the monthly heating degree-days from column 2 times 5.6 Btu/°F-day/ft² times the 2000 ft² floor area of the house. The monthly solar % given in the sixth column is simply column 5 divided by column 4, expressed as a percentage.

Albuquerque, NM					
Month	heating °F-days	Solar Input Btu /day/ft²	Heating Requirement Btus	Solar Heat Input (246 ft²) Btus	Solar %
Jan	955	1220	10,696,000	9,303,720	87.0%
Feb	700	1110	7,840,000	8,464,860	108.0%
Mar	561	800	6,283,200	6,100,800	97.1%
Apr	301	490	3,371,200	3,736,740	110.8%
May	89	360	996,800	2,745,360	275.4%
June	0				
July	0				
Aug	0				
Sept	18	650	201,600	4,956,900	2458.8%
Oct	259	1040	2,900,800	7,931,040	273.4%
Nov	621	1200	6,955,200	9,151,200	131.6%
Dec	921	1200	10,315,200	9,151,200	88.7%

Discussion of Results: The solar percentage results are quite high, indicating that the 246 ft² glazing area is more than required for this area of the country. Thus the calculations will be repeated with 170 ft² of south-facing glazing instead of 246 ft². The results are summarized in the table below. These results are much closer to the target level of 60% of heating requirement to be supplied by the solar system. There is more than enough solar input for the heating requirements in May, September and October. For the rest of the heating season, the solar percentage is between 57% and 86%.

28

| Albuquerque, NM | | | Heating | Solar Heat | |
Month	heating °F-days	Solar Input Btu /day/ft²	Requirement Btus	Input (160 ft²) Btus	Solar %
Jan	955	1220	10,696,000	6,051,200	56.6%
Feb	700	1110	7,840,000	5,505,600	70.2%
Mar	561	800	6,283,200	3,968,000	63.2%
Apr	301	490	3,371,200	2,430,400	72.1%
May	89	360	996,800	1,785,600	179.1%
June	0				
July	0				
Aug	0				
Sept	18	650	201,600	3,224,000	1599.2%
Oct	259	1040	2,900,800	5,158,400	177.8%
Nov	621	1200	6,955,200	5,952,000	85.6%
Dec	921	1200	10,315,200	5,952,000	57.7%

Example #7: To illustrate the effect of climate in a different part of the country repeat the last set of calculations for Kansas City, MO. Use 160 ft² of glazing, 2000 ft² floor area and 5.6 Btu/°F-day/ft² as the heat loss rate for the house.

NOTE: Table 6 below shows the average transmitted solar radiation rate through double glazing by month in Kansas City, MO and the heating degree days by month in Kansas City, MO. These figures came from page 117 (the Kansas City page) in the NREL *Solar Radiation Data Manual for Buildings.*

Kansas City, Missouri	Average Transmitted Solar Radiation (Btu/ft²/day) for Double Glazing, Uncertainty ± 9%						
	Jan	Feb	Mar	Apr	May	June	July
South Unshaded	880	850	760	640	520	470	500
South Shaded	860	790	610	420	350	350	350
HDD, Base 65°F	1218	946	691	325	135	7	0

Kansas City, Missouri	Average Transmitted Solar Radiation (Btu/ft²/day) for Double Glazing, Uncertainty ± 9%					
	Aug	Sept	Oct	Nov	Dec	Year
South Unshaded	610	760	920	840	780	710
South Shaded	380	550	810	810	760	590
HDD, Base 65°F	6	56	279	657	1073	5393

Table 6. Transmitted Solar Radiation Rate and HDD data for Kansas City, MO

Source for this data: NREL, *Solar Radiation Data Manual for Buildings*, p 117

Solution: The calculations are summarized in the table below. The data for heating degree-days and solar input through south-facing glazing came from page 117 of the NREL publication, *Solar Radiation Data Manual for Buildings* (see **Table 6** above). The other columns were calculated just the same as in **Example #6.**

Kansas City, MO Month	heating °F-days	Solar Input Btu /day/ft^2	Heating Requirement Btus	Solar Heat Input (200 ft^2) Btus	Solar %
Jan	1218	880	13,641,600	4,364,800	32.0%
Feb	946	850	10,595,200	4,216,000	39.8%
Mar	691	760	7,739,200	3,769,600	48.7%
Apr	325	640	3,640,000	3,174,400	87.2%
May	135	520	1,512,000	2,579,200	170.6%
June	7				
July	0				
Aug	6				
Sept	56	760	627,200	3,769,600	601.0%
Oct	279	920	3,124,800	4,563,200	146.0%
Nov	657	840	7,358,400	4,166,400	56.6%
Dec	1073	780	12,017,600	3,868,800	32.2%

As was the case for Albuquerque, there is more than enough solar input from 160 ft^2 of south-facing glazing to provide the heating requirements for May, September and October. For the rest of the heating season, however, the solar percentage is between 32% and 87% for Kansas City, as compared with 56% to 83% for Albuquerque. The solar percentage is significantly lower in Kansas City than in Albuquerque for all of the months except April. There are indeed less clouds in the sky in the southwestern U.S.

The glazing area determined by this method can be taken as an approximate value. More area or less area may be used to accommodate convenient sizes of components actually used for the passive solar heating system. In fact, Anderson & Wells (ref #2 under "Related Links & References" for this course), suggest that your answer to the question, "How large do you want the system to be?" is a common, direct, and useful method of determining size. The calculations illustrated above, however, provide a means of estimating solar % of heating requirement for a given size passive solar system, at a given location, with given building heat loss characteristics.

6. Choice of the Type(s) of Passive Solar System to Use

After deciding upon an approximate total area of glazing, it is necessary to choose from among the five basic passive solar heating systems discussed above, **direct gain, thermal storage wall, attached sunspace, thermal storage roof, and convective loop.** Any combination can be used to come up with the desired total area of glazing.

For retrofit on an existing home, the attached sunspace and the convective loop (solar chiminey) are especially suitable. Also, one may add more south-facing windows or upgrade existing south-facing windows to increase direct gain. If a thermal storage wall were to be used for retrofit, the choice would probably be made for a "water wall" rather than a masonry wall, because of the difficulty of inserting a masonry wall for retrofit.. A thermal storage roof (roof pond) is another possibility to go onto a flat roof or into an attic under glazing on a pitched roof. Remember, however, that a thermal storage roof requires a somewhat elaborate drainage system, movable insulation to cover and uncover the water at appropriate times, and a structural system to support up to 65 lbs/sq ft dead load.

For new construction, any combination of the five types of systems can be used. Maximizing southern exposure for the building and using a lot of south-facing windows would be typical in order to obtain direct gain of solar heat. A masonry thermal storage wall in front of some of the south-facing glazing, could readily be included for new construction. An attached sunspace works best if the back of the house faces south.

7. Sizing Solar Storage

Both of the references mentioned in the previous couple of chapters include guidelines for required quantity of thermal storage. Anderson & Wells, in *Passive Solar Energy: The Homeowners Guide to Natural Heating and Cooling*, (Reference #2) give a rule of thumb for thermal storage requirement as 2 cubic feet of concrete, brick or stone for each square foot of glazing, if the sun shines directly upon the storage material. If the sun heats air, which in turn heats the thermal storage material, then the requirement is given as four times that much, or 8 cubic feet of masonry storage per square foot of glazing.

Goswami, Krieth, & Krieder (Reference #1) give a rule of thumb that there should be 613 kJ/°C for each m^2 of glazing, if the sunlight shines directly on the storage material. They also give a requirement of four times that much if the sunlight does not shine directly on the storage mass. Converting units, this rule of thumb is equivalent to 30 Btu/°F for each ft^2 of glazing. This can be converted to the volume of material needed for thermal storage in ft^3 using the information in **Table 7**, below.

The volume of any of the materials in **Table 7**, needed per ft^2 of glazing can be found by dividing 30 Btu/°F by the volumetric heat capacity $(Btu/ft^3/°F)$ for that material.

Table 7. Thermal Properties of Some Materials

Material	Specific Heat (Btu/lb/°F)	Density lb/ft³	Volumetric Heat Capacity Btu/ft³/°F
Air (75°F)	0.24	0.075	0.018
Sand	0.191	94.6	18.1
White Pine	0.67	27	18.1
Gypsum	0.26	78	20.3
Adobe	0.24	106	25
White Oak	0.57	47	26.8
Concrete	0.2	140	28
Brick	0.21	140	28
Heavy Stone	0.21	180	38
Water	1	62.5	62.5

Example #8: What would be the storage volume requirement per ft^2 of glazing based on the guideline of 30 Btu/°F per ft^2 of glazing, for concrete, heavy stone, gypsum and water?

Solution: As noted above, the volume of storage needed for each material per ft^2 of glazing will be 30 Btu/°F/ft^2, divided by the volumetric heat capacity of that material. The results of those calculations are:

> **Concrete:** 30/28 = **1.1 ft³/ft²**
> **Heavy Stone:** 30/38 = **0.79 ft³/ft²**
> **Gypsum:** 30/20.3 = **1.5 ft³/ft²**
> **Water:** 30/62.4 = **0.48 ft³/ft²**

Example #9: What would be the volume of thermal storage needed for the 2000 ft² house with 170 ft² of south facing glazing, considered above, for each of the following thermal storage materials?

      a) Concrete
      b) Heavy Stone
      c) Water

Solution: Multiplying each of the results from Example #8 by the 170 ft² glazing area, gives the following thermal storage volume requirements:

    Concrete: $(1.1 \text{ ft}^3/\text{ft}^2)(170 \text{ ft}^2) = \textbf{187 ft}^3$
    Heavy Stone: $(0.79 \text{ ft}^3/\text{ft}^2)(170 \text{ ft}^2) = \textbf{134 ft}^3$
    Water: $(0.48 \text{ ft}^3/\text{ft}^2)(170 \text{ ft}^2) = \textbf{82 ft}^3$

As these examples illustrate, Anderson & Wells give a more conservative thermal storage requirement than Goswami, Krieth, & Krieder.

8. Summer Shading of Passive Solar Glazing

In order to reduce unwanted heat gain in the summer, it is helpful to provide shading above south-facing solar glazing. The NREL publication, *Solar Radiation Data Manual for Buildings*, provides recommendations for summertime shading of south-facing vertical passive solar glazing. Near the top of the page for each of the 239 stations represented in the publication, there is a figure similar to **Figure 8**, below, showing the length of overhang and height of overhang above the top of the glazing for each foot of vertical height of the glazing. The recommended shading geometry provides a balance between the need for maximum heat gain during the heating season without creating unreasonable heat gain during the cooling season. This can be done because the sun is much lower in the sky in the winter than in the summer, as illustrated in **Figure 9**, below.

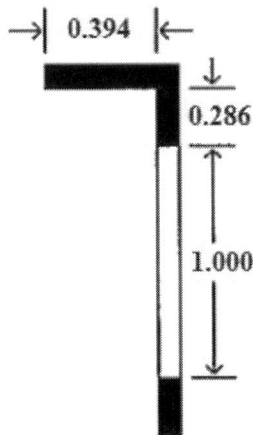

Figure 8 – NREL Recommended Shading Geometry for Albuquerque, NM

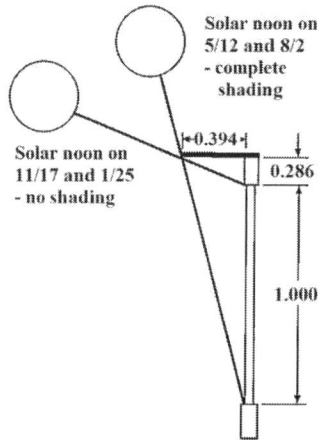

Solar noon on
5/12 and 8/2
- complete
shading

0.394

0.286

Solar noon on
11/17 and 1/25
- no shading

1.000

Figure 9 – Effect of Winter and Summer Solar Altitude Angle

Example #10: Determine the average transmitted solar radiation (Btu/ft2/day) for a double-glazed, south-facing vertical window with recommended shading and with no shading in January and in July for

a) Kansas City, MO

b) Albuquerque, NM

Solution: The following values can be found on the Kansas City, MO page of the NREL Data Manual for Buildings (see the data in **Table 5** above):

Results for Kansas City, MO:

	January	July
Shaded	880	500
Unshaded	860	350
% Decrease	2.3%	30%

Similarly, the following values can be found on the Albuquerque, NM page of the NREL Data Manual for Buildings (see the data in **Table 4** above):

Results for Albuquerque, NM:

	January	July
Shaded	1230	460
Unshaded	1220	350
% Decrease	0.8%	24%

As shown by these results, the NREL recommended shading only decreases the transmitted solar radiation slightly in the winter heating season, but decreases it significantly in the summer cooling season at both of these locations.

Example #11: An awning is to be used for shading a 24" x 16" south-facing window (24 " vertical measurement) in Albuquerque, NM. Determine the NREL recommended height of the awning above the top of the window and the distance that the awning should extend out from the window. The shading diagram from the NREL Manual page for Albuquerque is shown in Figure 8.

Solution: Using the numbers from the NREL diagram shown in **Figure 8** above:

Height of awning above window = (0.286)(2 ft) = **0.572 ft**

Distance to extend from wall = (0.394)(2 ft) = **0.799 ft**

9. NASA/Langley Surface Meteorology and Solar Energy Website

The NREL Passive Solar data manual that has been discussed above provides solar radiation data for the United States and its territories. The NASA Langley Surface Meteorology and Solar Energy website to be discussed here, however, gives access to over 200 satellite-derived meteorology and solar energy parameters for locations all over the world. The monthly averages in the NASA Langley data tables come from 22 years of data. Data for a site of interest can be accessed by entering the latitude and longitude of that site.

This worldwide data set can be accessed from the website: http://eosweb.larc.nasa.gov/sse/. Clicking on "Meteorology and Solar Energy" in the menu on the left side of the page will get you started. On the next page you should click on "Data table for a particular location," also in a menu on the left side of the page.

This will take you to a page with two boxes in which to enter the latitude and longitude of the site for which you want to obtain information. Note that 0 to 90 degrees south latitude should be entered as a negative number and 0 to 90 degrees north latitude should be entered as a positive number. For longitude entries, 0 to 180 degrees east is positive and 0 to 180 degrees west is negative. This information is also given below the latitude and longitude entry boxes on the website.

After clicking "submit" for your latitude and longitude entries or for your map location, a site will come up with a wide number of parameters to choose from. There are many parameters for you to choose from. You can click on the parameter or parameters for which you want values. Press Ctrl to select more than one parameter in the same category. After selecting the parameters for which you want values, you may click "Submit" at the bottom of the list. This will take you to a table or tables with the data that you requested.

The parameters from which you can choose for your site of interest are summarized in the outline below. The section, "Parameters for Tilted Solar Panels" contains data most similar to that in the NREL Passive Solar manual.

Parameters from which to Choose for Output:

Geometry
- Latitude and longitude (center and boundaries)

Parameters for Solar Cooking
- Average insolation
- Midday insolation
- Clear sky insolation
- Clear sky days

Parameters for Sizing and Pointing of Solar Panels and for Solar

Thermal Application

- Insolation on horizontal surface (Average, Min, Max)
- Diffuse radiation on horizontal surface (Average, Min, Max)
- Direct normal radiation (Average, Min, Max)
- Insolation at 3-hourly intervals

- Insolation clearness index, K (Average, Min, Max)
- Insolation normalized clearness index
- Clear sky insolation
- Clear sky insolation clearness index
- Clear sky insolation normalized clearness index
- Downward Longwave Radiative Flux

Solar Geometry
- Solar Noon
- Daylight Hours
- Daylight average of hourly cosine solar zenith angles
- Cosine solar zenith angle at mid-time between sunrise and solar noon
- Declination
- Sunset Hour Angle
- Maximum solar angle relative to the horizon
- Hourly solar angles relative to the horizon
- Hourly solar azimuth angles

Parameters for Tilted Solar Panels
- Radiation on equator-pointed tilted surfaces
- Minimum radiation for equator-pointed tilted surfaces
- Maximum radiation for equator-pointed tilted surfaces

Parameters for Sizing Battery or other Energy-storage Systems
- Minimum available insolation as % of average values over consecutive-day period
- Horizontal surface deficits below expected values over consecutive-day period
- Equivalent number of NO-SUN days over consecutive-day period

Parameters for Sizing Surplus-product Storage Systems
- Available surplus as % of average values over consecutive-day period

Diurnal Cloud Information
- Daylight cloud amount
- Cloud amount at 3-hourly intervals
- Frequency of cloud amount at 3-hourly intervals

Meteorlogy (Temperature)

- Air Temperature at 10
- Daily Temperature Range at 10 m
- Cooling Degree Days above 18°C
- Heating Degree Days below 18°C
- Arctic Heating Degree Days below 10°C
- Arctic Heating Degree Days below 0°C
- Earth Skin Temperature
- Daily Mean Earth Temperature (Min, Max, Amplitude)
- Frost Days
- Dew/Frost Point Temperature at 10 m

Meteorology (Wind)

- Wind Speed at 50 m (Average, Min, Max)
- Percent of time for ranges of Wind speed at 50 m
- Wind Speed at 50 m for 3-hourly intervals
- Wind Direction at 50 m
- Wind Direction at 50 m for 3-hourly intervals
- Wind speed at 10 m for terrain similar to airports

Meteorology (Wind): be sure to select an appropriate VEGETATION type(s)
1. Percent difference for Wind Speed at 10 m from ave. Wind Speed at 50 m

2. Gipe Power Law used to adjust Wind Speed at 50 m to other heights
3. Wind Speed at 50, 100, 150, and 300 m
4. Wind Speed for several vegetation and surface type

- Vegetation (up to 17 types of vegetation/surface can be selected)
- Height (_____) Minimum/Maximum heights are 10 and 300
  (Choose a vegetation/surface for 1, 2, & 3. Choose a height for 2 & 4

Meteorology (Other)

- Relative Humidity
- Humidity Ratio
- Atmospheric Pressure
- Total Column Precipitable Water
- Precipitation

Supporting Information
- Top-of-Atmosphere insolation
- Surface Albedo

References

1. Goswami, D. Y., Krieth, Frank, and Kreider, Jan F., *Principles of Solar Engineering,* Philadelphia: Taylor & Francis, 2000.

2. Anderson, Bruce & Wells, Malcolm, *Passive Solar Energy: The Homeowners Guide to Natural Heating and Cooling,* Andover MA: Brickhouse Publishing Co., 1981 (available for free download at the website given below:)
http://www.builditsolar.com/Projects/SolarHomes/PasSolEnergyBk/PSEbook.htm

3. Bengtson, H. H., "Principles of Passive Solar Heating Systems and How They Work," BrightHub.com, 2010

4. Bengtson, H. H., "Estimating Solar Radiation Rate to the Tilted Surface of a Solar Panel in the U.S.," Brighthub.com

5. National Renewable Energy Laboratory (NREL), Solar Radiation Data Manual for Buildings, available for free download at:
http:rredc.nrel.gov/solar/pubs/bluebook/.

6. NASA/Langley Surface Meteorology and Solar Energy website, available at:
http://eosweb.larc.nasa.gov/sse/

41306355R00028

Printed in Poland
by Amazon Fulfillment
Poland Sp. z o.o., Wrocław